THE GREAT CA$H HUNT

33 WAYS

To Boost Cash Flow
in Your Kitchen and Bathroom Business

By Stephen P. Vlachos, CKD, CBD and Leslie L. Vlachos, M. Ed.

Published by the National Kitchen and Bath Association

BOOK STORE

- **Publisher: Nick Geragi, CKD, CBD, NCIDQ**

- **Director of Communications: Donna M. Luzzo**

- **Director of Marketing: Nora DePalma**

- **Executive Director: Paul A. Kohmescher, CAE**

The National Kitchen & Bath Association (NKBA) is the leading international organization exclusively serving the kitchen and bathroom industry. NKBA is dedicated to researching and providing information on all facets of kitchens and bathrooms, and will continue to pursue timely subjects that affect the industry and those working in it.

ISBN 1887127-43-7

Information about this book and other NKBA publications, membership and educational seminars may be obtained from the National Kitchen & Bath Association, 687 Willow Grove Street, Hackettstown, NJ 07840; phone 800-THE-NKBA; fax 908-852-1695; e-mail educate@nkba.org.

This book is intended for professional use by kitchen and bathroom business professionals. The procedures and advice herein have been shown to be appropriate for the applications described, however, no warranty (expressed or implied) is intended or given. Moreover, the user of this book is cautioned to be familiar with business management and accounting principles.

OTHER CONTRIBUTORS

- Editor: Janice Von Brook
Janice Lamb, Office Services

- Content Reviewers: Paul Kohmescher, CAE, Executive Director
National Kitchen & Bath Association

Nick Geragi, CKD, CBD, NCIDQ, Director
Education and Product Development
National Kitchen & Bath Association

- Electronic Publisher
and Designer: Janice Von Brook
Janice Lamb, Office Services

FROM THE PUBLISHER

Dear Industry Professional:

How liquid are you? If you had to find cash in your business this very moment, could you? Take this quick test. Divide your cash-on-hand by your current liabilities. Are you solvent? Without cash, everything you do in business comes to a screeching halt. Without cash, you can't pay your suppliers, your staff or tradespeople, the utility expenses, or even yourself. Sure, many businesses operate from job to job and from month to month using the next project to pay for the last one. But how long can a business truly survive this way before failing?

If you have ever looked at the sales produced by your company and wondered why more cash isn't sticking to your pocket, read *The Great Cash Hunt*. This NKBA publication profiles many of the practical "nuts-and-bolts," cash-producing strategies that Stephen and Leslie Vlachos have used to create a financial turnaround at Atlantic Kitchen Center. You'll find their creative ideas for raising cash and controlling expenses to be so applicable to your business, you'll think they're talking about you.

Nick Geragi, CKD, CBD, NCIDQ
Publisher, NKBA Books
Director of Education and Product Development

NKBA'S ELEMENTS OF SUCCESS

Earth, fire, water and air are the elements necessary to sustain life. In similar fashion, a business needs a solid Start-up, successful Management, Profit and good Business information to thrive. NKBA recognizes these basic business needs and offers this new three-book series which focuses on the elements of success essential to your kitchen and bathroom firm.

Managing Your Kitchen and Bathroom Firm's Finances for Profit   by Don Quigley
The Element of Water

Just as you can't survive without water, your business can't survive without profit. Profit should flow through your business, buoying up the equipment, the people and the services you provide. The key is to plan, and to work with your financial records to keep the profit reservoir full. If you don't, profit will slow to a trickle, and your business will eventually fail. Use this guide to help keep profits flowing in your business.

The Complete Business Management Guide for Kitchen and Bath Professionals ...
Starting and Staying in Business   by Hank Darlington
The Element of Fire

Fire provides the warmth that keeps us alive, the energy that moves us and powers machines, and a source of light when things get dark. So does innovative management propel a business forward and ensure that it continues on its path. Good management can lead a company through good and bad times. While poor management, like an untended fire, destroys. Learn to manage every area of your company to keep projects moving, employees energetic and your business HOT!

The Kitchen and Bath Dealer Operating and Performance Survey   by NKBA
The Element of Air

Our intake of air is essential to life. For a company, the intake and use of business information is essential for survival. This publication will provide important financial and management comparisons for kitchen/bath businesses. You'll be able to assess your own business practices and make modifications accordingly to stabilize your business atmosphere. A must-have for every business. Available to NKBA members only.

CONTENTS

INTRODUCTION

Our kitchen and bathroom business started on "a wing and a prayer" in 1983. We had good ideas. We had enthusiasm. We inspired and encouraged each other. Having gained management experience with other big companies, we had the youthful arrogance of thinking that our ideas were better than the next guy's. We also had the energy needed to put our ideas into action. In short, we had the stuff it takes to become an entrepreneur. And so we did!

Our business was part of the generation of kitchen and bathroom dealers who lived through the easy spending of the 1980s. Customers came into our showroom with their checkbooks already open. Their only question was, "How much do I make the check out for?" Often, we would shrug our shoulders and come up with a figure that sounded reasonably thought out. One particular sunny Saturday afternoon holds fond memories. Our business had just closed for the week and we were relaxing at a cafe table overlooking Portland's scenic harbor. We were basking in the glow of the warm Maine sunshine, the glow brought on by some chilled Chardonnay and the very heady glow of cash flow. Our discussion focused on the number of jobs we had sold that week. We guessed that there had to be 12 deposit checks stuffed into the top drawer of our desk. Neither of us knew how much cash we had collected, only that there was a lot of it. We were in no rush to leave the cafe. We were in no rush to get those checks into the bank. We were high on self-employment. Life was good.

As long as cash is flowing, most of us consider ourselves to be good business people. Stories abound, however, of businesses starting up and then going out of business. There is a general assumption that if our doors are open, we must know what we're doing. Certainly, our business skills must be better than the guy down the street who just closed up shop. Or are we just lucky?

Since that day in the early 1980s, we have become business people. We now know exactly how many deposits we receive each week. We know exactly what our daily cash position is. We know exactly what we owe and what we are owed. Most importantly, we know the difference between cash flow and true profitability. And we know that our business, and your business, need both to survive and achieve prosperity.

In 1987, a fortuitous thing happened. A gentleman walked into our showroom and offered to buy our business. We were quite surprised as we had never considered the possibility of selling the business. At that time it had not even occurred to us that we had created something that had value to anyone else. But we had. And sell our business we did. We sold our kitchen and bathroom business rather profitably with zero debt.

Sounds like a fairy tale doesn't it? Well, the unbelievable recession of the late 1980s (or was it a depression?) raised its ugly head and changed it from a fairy tale to a scary tale. In three years the buyers of our business accumulated over $400,000 in debt. Needless to say, their cash flow was non-existent. We, on the other hand, were off creating other businesses, and trying to reachieve the cash flow we had once enjoyed. After we sold our kitchen and bathroom business, we came to truly understand the dynamics of personal cash flow. Once we were without monies generated weekly by our former business, we could not patch the leaks as our recently ballooned savings account quickly lost air.

We decided to try and save our former business. We let the buyers walk away from the mess they created. We charged in full of plans to save the business we had started. We knew the industry. We knew how to design. We knew how to sell. We were ready to reclaim our cash cow. But the cow had been milked dry.

We struggled to keep the business afloat. Staff was laid off. Debt repayment plans were negotiated with banks and vendors. We had been gone only three years and couldn't believe how much the climate had changed. The consumers had changed. The market had changed. There were 15 other places in town where a homeowner could buy a kitchen. Each showroom was loaded with cabinet lines in an attempt to service every conceivable price point that existed. Our showroom was no different. Price was now the buyer's top issue.

We began a hunt to find new ways to generate cash for our business. The true costs of being in business were painfully identified. Gross margins were worked and reworked. Displays were sold off to raise cash. We sold our delivery truck. Every item that was not nailed down was sent back to suppliers for credit. We cut out phone lines and eliminated the cleaning service. Every aspect of our business was examined in an effort to find more cash.

We knew that we were talented kitchen designers and productive salespeople. But, more importantly, we knew that we had to become even more effective business people if we were to have any chance of surviving. Strategies were developed and implemented to find and capture every dollar possible. And we did it! The cash did begin to flow. We met all of our obligations to repay the vendor debt accumulated by the former owners. Even now, as we are well past the survival stage, we continue to hunt for cash while providing our clients with quality projects. Everything we do is with a keener eye on the bottom line.

In the kitchen and bathroom industry, conversation at meetings and gatherings tends to focus on the design and dollar size of projects that are being worked on. Seldom do we hear anyone talking about the profitability of their projects. As our industry colleagues wax eloquently about the work they do, there is always an unasked question, "Did you make any money?" Unfortunately, many of us wouldn't know the answer to that question. Further, many of us are uncertain how to find out. This book shares with you some of the strategies we have used to determine our true profitability and to boost our cash flow. We hope that we can help you in your own *Great Cash Hunt.*

1 TWO PERCENT IS HEAVEN SENT!

The most important cash that we can boost is the cash that flows through to **you**, the business owner. After all, if there is no money for you at the end of the day then you are going to be very unhappy. It is doubtful that you'll be hiring any new staff, enlarging your showroom or buying a new car. If you're not enjoying personal cash, why put up with the headaches of being a business owner? If you're not generating personal cash, it's a good bet that your growth rate will be as flat as your wallet. Remember, profits are good, and cash is paramount.

It is often difficult to understand the true profitability of a small, closely held business. Bottom-line net profit often ranges from two percent to five percent. Anything more than that probably means you're not paying yourself enough. Anything less may jeopardize bank financing and supplier credit arrangements. It has always been our goal to show a profit somewhere between two percent and five percent. If we generate additional profit, then we try to spend monies on goods and services that will improve our business. Most folks would agree that this a better way to spur the economy in the United States than sending that money in as a tax payment.

Let's assume that your firm is generating two percent or more net profit. (If it isn't, then it will by the time you finish reading this book.) At the end of the year, are you going to look at your bottom line and wonder where the cash went? It happens to business owners all the time! The answer to that quandary is to take **your** cash right up front.

Open an interest-bearing account at the financial institution of your choice. Then **every** day take two percent of **every** dollar that your firm receives and put it into that account. Make sure that only the business owner can access the funds that accrue. In our firm, we call it the *Special Account*, and it appears as an entry on our *balance sheet*. At the end of the year you will no longer be wondering where your two percent profit went. It will be right there—earning interest!

Let's assume that your business had $800,000 in sales this year:

$$.02 \times \$800,000 = \$16,000$$

If your business grows by five percent each year for five years;

$$.02 \times \$800,000 = \$16,000$$
$$.02 \times \$840,000 = \$16,800$$
$$.02 \times \$882,000 = \$17,640$$
$$.02 \times \$926,100 = \$18,522$$
$$.02 \times \$972,405 = \$19,448$$

The total two percent saved up-front in five years is $88,410, plus interest!

You already **know** that you're going to generate two percent or more net profit this year. Taking the two percent up front ensures you that cash will be there for you when you need it. Use it to plan your retirement, pay college tuition or buy your own building. Best of all, when you have a tough day, and we all have plenty of those in our business, it's gratifying to check out **your** two percent account and find that it's bigger than it was the day before.

The transfer of the two percent funds does not have to be a laborious project. You don't need to sit down and write checks. Instead, prepare a one-page fax sheet with your bank's name on it and use it every day that you get some cash in. It could look as simple as this:

To: **Armadillo Bank & Trust**
Fax Number: **555-1212**

From: **Dynamic Kitchen Center**
Fax Number: **555-1313**

Date: _____

Please transfer $ _____
from Account # _____
to Account # _____

Name _____

Signature _____

It is a very simple process! Finally, if you think you can't afford to set aside two percent, then it's time to question the viability of your business.

2 GOOD NEWS? HIGHER MARGINS MEAN LESS SALES!

Many of us have an ingrained belief that more sales mean more profit. But that's not always the case. There have been countless companies that have had to close their doors because their selling success outpaced their success in producing an adequate cash flow.

In our industry, this is especially true with stocking retailers and those whom we call *whotailers* (a combination of wholesalers and retailers). If business is booming, more inventory is required (more trucking, more salespeople, more office help). The ability to turn receivables into cash quickly is absolutely critical. Even if you are lucky enough to have great customers that pay in 30 days, you may still be in trouble. Your biggest supplier will probably offer great discounts if you pay within 10 days. It would seem clear that taking that discount would help ensure profitability in your venture. But how are you going to come up with the cash in 10 days if your customers are paying in 30? How are you going to pay for the increased inventory? Or the additional sales and support help?

Every year we hear about companies that are successful in selling great numbers of products, but have to close up shop. Inevitably, it's because they didn't understand the art of cash flow. In the stocking end of our industry, increased sales are necessary to have increased profits. However, you can never lose sight of the fact that cash is paramount!

With kitchen and bathroom showrooms, the situation is often very different. Most of us will get a deposit at the time of signing a sales agreement. If you are not getting a deposit, you really ought to reconsider the way you're doing business. Deposit amounts should be equal to the profit you hope to make on the job. This way, you're receiving your profit right up front. Of course, you're liable for those customer deposits until the job is complete. So how do you increase profits? One way is to increase your sales. Unfortunately, with many small, closely held businesses that means that the owner now has to work 16-hour days instead of the normal 12! Is there another way? Although it may seem sacrilegious, the best way may be to raise your margins and lose some sales. In fact, increased margins with lower sales usually result in better profits.

Let's say that you decide to increase your margin by five percent. The increase may cause you to lose five percent of the jobs you're working on. The effect on profits can be calculated as follows:

PRESENT		ADJUSTED	
Sales	$1,000,000	Sales	$950,000
Gross Margin	35%	Gross Margin	40%
Gross Profit	$ 350,000	Gross Profit	$380,000

If you usually sell based on price only, you may have to rework the calculation above to reflect a greater drop in sales. If, however, your sales effort is based on quality design, quality products and quality follow-through, you will have no problem raising your selling margins. Best of all, you will increase your profits without increasing your overhead. With fewer jobs, you will have fewer headaches. Fewer headaches mean more enjoyment. You might even be able to cut your hours. At any rate, you owe it to yourself to seriously explore raising your profits by raising your margins.

3 INVENTORY IS LIKE PURGATORY!

Purgatory is somewhere between heaven and hell. For a custom-kitchen dealer, having inventory is much the same. It's like sitting around waiting for money to happen.

Why do we have inventory? And we all do! Probably because we made mistakes on some jobs. Also, we were nice enough to take back a few items from customers, a cutlery tray here, a sink strainer there. Throw in a few B-24s along with some damaged countertops and we're talking about thousands of dollars! Whose money is that anyway? It's yours! It doesn't help anyone to tie up cash like that. Besides, how many times have you, or worse, someone you were paying, had to move that stuff around so that you could get legitimate products in and out of your storage space? So how do we turn this inventory into cash?

1. If any of this inventory is resalable, then your first step should be to try and send it back to suppliers in lieu of payment. A friendly arm-twisting of your supplier's sales representative will often help you get things returned. Even if the supplier doesn't allow you full credit, you will be way ahead of any other method of disposing of it.

2. Donate it! Find a local charity and give it all the material it will take, making sure that you get a signed receipt. Your charitable contribution can then be figured at full retail price. Multiplying your tax rate (approximately 30 percent) times the full retail price will help you figure the true compensation you are receiving for the items. At any rate, you will receive more for

your inventory than you would if you had a yard sale. Over the years we have donated various items to Habitat for Humanity, the Salvation Army and Goodwill Industries. All were very pleased to have the materials, and the donations best suited both of our needs.

3. Have a yard sale. This is the third option because you really don't want to do this. You have to lug all the stuff out for people to see. You have to endure all kinds of inane questions and comments. Then you have to lug half of it back. All for pennies on the dollar. The best way to boost cash is to work harder on the first two options.

The best way to keep track of what comes back into your warehouse is to make sure that all items are signed in. That can be accomplished with a simple form such as the following. If you use it conscientiously, you will find that it expedites items to be returned to vendors for credit. You will also benefit from fewer "lost" items. Further, you will be developing values to be used in your next physical inventory.

ITEM FOR WAREHOUSE INVENTORY

Job Name _____   **Date** _____

Vendor _____

Item Description _____

_____

_____

Reason for Return _____

_____

Returnable to Vendor for Credit? _____

Adding Item to Inventory? _____

Anticipated Vendor Credit Amount $ _____

Item Cost Including Freight $ _____

Item Retail Value $ _____

Signature _____

It works well for us to have an inventory report. Basically, it is a listing of items returned to our warehouse for any reason. Our staff is encouraged to look first at the list before placing any new orders. We allow them to list any warehouse item as having no cost on their job orders. That way it helps them increase their commission base while helping us turn lost dollars into cash.

4 DISPLACE DECEASED DISPLAYS!

Over the years we have visited countless showrooms. Inevitably, there is at least one display that is obsolete. Often times, a product no longer offered by the dealer is on display. Sometimes dealers will tell us that they have not changed the display because they're too busy. More often, however, they have not changed the display because they're lacking the cash needed for a new one. Think about that, not only is there money tied up in that display, there is also no new cash being generated by these items taking up valuable space in the showroom.

The first thing that should happen is to get rid of the offending display. Selling it could produce some quick cash. Donating it to a non-profit group will produce some down-the-road benefits. Donating it to a non-profit theater group could have both long-term tax benefits and immediate marketing benefits if you make it part of the arrangement that you get mentioned prominently in their event programs. You can get the same tax benefits and possibly some publicity by donating your displays to Habitat for Humanity, public television, your local NARI (National Association of the Remodeling Industry) or NAHB (National Association of Home Builders) fundraiser. Goodwill Industries and the Salvation Army have also been pleased to receive our obsolete displays.

If cash is tight, buy some affordable wallpaper to cover the area of the old display. Then take photographs of kitchens you've done and blow them up to 8" x 10". Matte them and hang them attractively on the wallpapered wall. If you don't have any photographs (you should!), then contact your cabinet supplier. Most will have quality pictures available for your use.

Your new kitchen and bathroom gallery will generate more new sales opportunities than your obsolete display ever could. Add some plants and signage and you may choose never to hang a cabinet display there again!

5 GOOD RIDDANCE TO NON-PERFORMERS!

For whatever reason, most of us end up displaying products that we don't sell. If we choose not to sell appliances, why are there cooktops and ovens prominently displayed? If we do not actively sell ceramic tile, why do we choose it as a display surface?

The old rule was, "If you don't sell it, don't show it." That should be revised to say, "If you paid for it and don't sell it, don't show it." There is no reason to go out and buy a cooktop for your display if you have no intention of being a cooktop retailer. If you feel that it is important to show that you know how to handle cooktops, use your photo enlargements as a demonstration. You may want to find an appliance distributor or retailer who will provide appliances on consignment. Our showroom, for example, is outfitted with high-end appliances made available by a distributor who understands that it is dealers, like us, who specify their products in our designs. When a high-enough margin is attainable, we may choose to sell the appliances ourselves. Usually, however, we will refer the client to a local appliance store who also works with the same distributor. Our *partnership* with this distributor allows us to turn a negative return on investment into a positive one. All other unsalable display items have been turned into cash.

If, for instance, granite countertops are too competitive for you to make a profit, tell your stone sources that you're taking them out. It should be worth it for one of your stone suppliers to provide you with a no-cost display in order to gain your exclusive referrals. The same holds true for sinks, faucets, tubs, toilets, hardware—anything that does not offer a financial return to you. If you take the time to understand exactly what each square foot of showroom floor space costs you each year, you will clearly understand the negative impact of non-producing display items. Take a look at this example using a built-in refrigerator with front and side panels:

Square Footage	Rent Per Square Foot	Appliance Cost	Panels	Total Investment
6 (3' x 2')	$576 (6 x $8 x 12 months)	$2200	$580	$3356

A built-in refrigerator is a wonderful product. But you're going to have to make a commitment to sell a lot of them to achieve any significant return on your investment. Don't forget that you will probably have to send two or more people to deliver this item, and you will probably also need to invest in an automated-type of tailgate for your truck. Adding up the cost of the appliance, including its panels and your delivery costs, will give you your true costs of putting this item on display. When you factor in the low-gross margin that you'll be able to place on this product, you may find that you will not be able to sell enough of them to have any sort of return on your investment. Concentrating instead on the use of creative cabinetry in your displays will cost you less and put more money in your pocket (where it's supposed to be!).

Think about this...

There are certainly a lot of changes on the horizon for our industry. The big home centers will get bigger and, most certainly, better. As they get larger and spread across the country, they'll squeeze out a good many kitchen and bathroom dealers who will try to compete with them on price. The cabinet manufacturers that supply these "mega" stores will no doubt grow along with them. But what happens to all those other manufacturers who then find their dealer-base shrinking? They'll have no choice but to find another way to market competitively. How? Possibly by cutting out the dealer.

Does that mean that kitchen and bathroom dealers are doomed? Not necessarily. But it certainly means that dealers have to anticipate and embrace change. The smartest move might be to understand and commit to *partnering*.

Partnering is when all parties involved in the sales and manufacturing process work together to service and please the end consumer. Traditionally, manufacturers have viewed the dealers as their customers. In fact, many manufacturers have had little idea what it took for the dealer to make a sale. They have felt that all they had to do was build cabinets, per the dealer's order, then drop them off and get paid. The dealers, on the other hand, have been quick to jump from supplier to supplier trying to find the best price. Neither party is going to make it in the future without understanding and fully committing their resources to each other.

Dealers should not be expected to carry the total burden of displaying manufacturers' products in their showrooms. Manufacturers, in return, should have the right to demand that fewer competitor products be found on the dealers' showroom floors. The sole goal should be to attract, sell and service the end consumer.

Just look at the *partnerships* being created in the automobile industry. Ford/Mercury, for example, is cutting out a number of their dealerships. The result is that the dealers who remain will enjoy increased market territory and, of course, less competition which drives their margins down. In order for a dealer to be a *partner* in this new scenario, he or she will have to cut down the number of car lines carried. It's this same win/win situation that will have to be employed by many in the kitchen and bathroom industry, if we hope to be around to see the year 2000.

Historically, dealers would take a quick scan of the local Yellow Pages to see who their competitors were. In the very near future, however, competition will come from every angle conceivable. Manufacturers might be selling direct on one side of town, while distributors are selling direct on the other. The end consumer might even purchase his or her kitchen or bathroom products from a faceless entity on the Internet.

If you don't have the resources to compete with the big home centers, then you need to start developing *partnerships*. They make sense. You need the combined strength of all your *partners*. Plan on celebrating your success, not lamenting your demise as the new century rolls in.

DON'T PAY FOR THAT NEW DISPLAY!

Times have changed. Dealers know it is no longer wise or prudent to load their showrooms with displays from large varieties of cabinet and countertop manufacturers. Neither is it prudent for manufacturers to allow their product to become one of many on display. The new concept is called *partnering,* and, it can, if you commit to it fully, produce more profit dollars for both dealer and manufacturer.

The key to holding on to more of your cash is to choose a very limited amount of suppliers to buy from. No matter how much you like the sales representatives or the delivery drivers from certain companies, the selection of suppliers you do business with should be based solely on profit potential. Although personal relationships in business are important, accruing cash for college tuition, vacations with your family and a stress-free retirement have to be your primary goal.

Once you've decided on your narrow list of manufacturers to represent, it's time to sit down with them and put together a display program. It's not enough to accept an extra 10 percent to 20 percent discount on your display order. All that means is that the manufacturer has covered costs while you lay out thousands and thousands of dollars to make the display look good. If a cabinet supplier insists on a program like this, it's time to find a new supplier!

Instead, try this conversation with your manufacturer, for a win/win situation. "We are committing our showroom and selling resources to a smaller number of suppliers, and we would like you to be one of them. Our goal is to increase the exposure and subsequent sales volume of your product which will benefit both of us. Redesigning our showroom to accomplish this goal will necessitate approximately $10,000, at cost, of new cabinetry from your firm. We suggest that the $10,000 be memo-billed for a period of 12 months from the date the new displays are installed. If we reach $100,000 in purchases over that 12 months, the bill will be considered paid in full. If we fall short of that sales goal, we will pay you whatever percentage we fall short, i.e., $90,000 in purchases will result in a payment of 10 percent of the $10,000." Again, if your supplier balks at a no-cash, up-front plan, then it's time to talk with someone else who appreciates what you can offer. A true *partnership* calls for both sides to contribute equally for the good of each other.

It has been our experience that most cabinet manufacturers have no idea what it takes for a retailer to make their product look good in a showroom. Many feel that a small extra discount is all it takes. We suggest that you estimate your total expenses for every new display, for review with your cabinet supplier. In that estimate, make sure you include the square foot share of rent and taxes to be borne by that display. Also include anticipated labor costs to hang the cabinetry, and for countertops, sinks, faucets, cooktop, hardware, paint, wallpaper and decorative accessories. It is important to let your *partner* know just what it takes to make the product attractive to the consumer client.

Here is a breakdown of a recent display installed in a dealer's showroom who did not *partner* with his suppliers:

TYPICAL DISPLAY	INVESTMENT
Cabinetry	$4,218
Solid-surfacing	1,771
Hardware	90
Sink and faucet	535
Labor	675
Painting	210
Lighting	112
Electrical	120
Ceramic tile	295
Dishwasher	544
Accessories	390

TOTAL $8,960

You also have to consider that this particular display is one of 12 in the showroom. We then divide the display's yearly share of showroom rental costs in order to help us understand how much business this display needs to generate to bring a return on our investment. To be completely accurate with your return-on-investment estimate, you should figure in the display's portion of heating, lighting, etc., but that gets rather cumbersome. In this particular case, this display and its share of rental burden look like this:

Display costs	**$ 8,960**
1/12 of showroom yearly rent **($2800/month x 12 months ÷ by 12 displays)**	**2,800**
Total cost for display this year	**$11,760**

As you can see, even with making an agreeable arrangement with your cabinet supplier to defray those costs, the dealer is still left with a substantial investment to make. Of course, to make this program work correctly for you, you'll need to make the same arrangements with your sink supplier, countertop provider, etc. The key, again, is to work with a more limited group of manufacturers and distributors. Make a commitment to them and expect their commitment in return.

To finish off your new display, you may want to seek out some allied retailers in your area. Is there a local kitchen or bathroom accessory store that would like to feature their products as part of your display? We've had no problem in obtaining ceramic tile, appliances, cooking accessories and so forth, from other retailers at no cost, in return for the exposure and referrals that we can offer.

A dealer who *partners* with his or her suppliers is a dealer who should expect better results. Your staff will not have to learn so many lines which should lessen mistakes. Your deliveries in and out will become fewer and more orderly. Your customers will be less confused as to what they have actually purchased. There will be fewer invoices to handle and pay. And, most importantly, you will be keeping the cash that you traditionally spent on new displays. Make the commitment!

7 AVERT AN IDENTITY CRISIS!

When you really think about it, most customers are buying you, not the cabinet lines that you represent. A year after they have purchased a new kitchen, very few have any idea who manufactured their cabinetry. They all, however, remember where they bought it. Why, then, spend an enormous amount of money on supplier brochures? Most consumers ignore the technical material in these brochures and simply use them as pictures of kitchens. You may be spending a great deal of money providing these pictures of kitchens to people who may never become your customer.

We suggest that you investigate developing your own brochure. One that details who you are and what you do. It should also explain why you are the best at what you do. We have found that some suppliers are willing to expend co-op dollars to help you develop your own presentation. In return, you may have to (or want to) put their logo somewhere on your brochure. You may also want to consider putting a private label on your cabinetry offering. Many manufacturers have become enlightened to the fact that it is you who sells their products, not them. With that in mind, they are more than willing to put your name on the cabinetry they build.

When a customer leaves your showroom after their first visit, they should have only your brochure. Once they become a serious client you may choose to provide them with manufacturer brochures and samples. Until that time, avoid an identity crisis and keep more cash in your checkbook!

 LET CUSTOMERS INCREASE YOUR CASH FLOW

Cash flow, at its best, is making sure that money that comes in exceeds money that goes out. The extra money left over is called profit. It sounds so easy; how come everybody can't generate profit? *Profit-eaters* can leap out and gobble up dollars while you're looking the other way. You work hard to profitably sell a client, and everything looks rosy until a staff member "mis-orders" something, your manufacturer builds an item incorrectly, your delivery person drops a cabinet or the installer on-site decides he's a better designer than you and changes some things. All *profit-eaters* can blind side you. What can be done? Well, you should check and double check your plan, your calculations, your orders. Have different members of your firm responsible for checking each other's work. This will help keep the *profit eaters* at bay. But human nature tells us that we will never be rid of them entirely. The next best thing to do is make sure that your cash flow is strong enough to get you through the potholes lining the road to prosperity.

The customer, of course, is the one with the cash. It's your job to make sure you get the cash from your customer. First of all you need to require a deposit with every order. We suggest that the deposit amount be equal to, but not more than, the profit you hope to generate on the project. If you think that you will gross a 40 percent profit, then require a 40 percent deposit. Make sure to get the balance on delivery or, if you're installing, get all but 10 percent. Controlling the money in this manner will keep the cash turning and allow you to do a better job for your customer. This is especially true when the *profit eaters* attack.

Another cash-smart move is to arrange an attractive prepay discount with your suppliers. If you can get an additional five percent by prepaying, you open up some options. If you have the cash available, go ahead and prepay the supplier when placing the order. Your gross margin just went up five points. If, however, you're like most of us who don't have extra cash lying around, offer an additional five percent to your customer for their prepayment. You may be surprised by how many customers will take you up on it. You won't make any additional margin, but you will be controlling all of the cash right up front.

During recessions, profit becomes almost secondary. Cash flow is the primary goal. During economic upswings you should have both profits *and* cash flow. Properly managing cash is an art that you need to perfect. If you're too busy to tend to cash flow, you may soon find yourself very *unbusy* with no business to tend to.

WEEKLY CASH FLOW REPORT

Week Ending:_____ **Total Cash Received $**_____ .

Customer Deposits Received: $ _____
   Customer Name _____ **Deposit Amount** _____
   Customer Name _____ **Deposit Amount** _____
   Customer Name _____ **Deposit Amount** _____
   Customer Name _____ **Deposit Amount** _____
   Customer Name _____ **Deposit Amount** _____
   Customer Name _____ **Deposit Amount** _____

Customer Payments Received: $ _____
   Customer Name _____ **Payment Amount** _____
   Customer Name _____ **Payment Amount** _____
   Customer Name _____ **Payment Amount** _____
   Customer Name _____ **Payment Amount** _____
   Customer Name _____ **Payment Amount** _____
   Customer Name _____ **Payment Amount** _____

Total Booked Sales: _____

Sales Tax Liability: _____

TOTAL AVAILABLE CASH: _____

     Checking Account Balance   **$** _____
     Money Market Account   **$** _____
     Special 2% Account   **$** _____
     Line of Credit Available   **$** _____

Cash Discounts Realized This Week: _____

Customer Deposits Anticipated Next Week: _____

Customer Payments Anticipated Next Week: _____

Anticipated Payroll Expense Next Week: _____

9 PROMPT PAY IS THE ONLY WAY

At the end of the year, most privately held companies will show a net profit somewhere between two percent and five percent. If you show less than that, you may have trouble getting financing. If you show more than that, then you are probably giving away too much in taxes.

Coincidentally, the two percent to five percent net profit is equal to the prompt-pay discounts offered by most suppliers. If you have not bothered with prompt-pay discounts, here is your chance to double your bottom line.

We know that taking every cash discount is easier said than done. But, making the commitment to try is a solid first step to boosting your cash. Just think about it. If your business purchased $500,000 worth of materials, with an average prompt-pay discount of three percent, that's a total of $15,000 of found cash. Think of all the things you could do with an extra $15,000!

It is important to start by charting the discounts available to you by every single supplier. We have found many times that suppliers, especially smaller ones, do not put discount terms anywhere on their invoices. They are hoping that you just won't ask. Make sure you do! Then make it standard procedure to pay those with the biggest discounts first. If someone doesn't offer any type of prompt-pay incentive, then don't pay until 30 days at the earliest! It is worthwhile to call those suppliers not offering prompt-pay discounts and explain to them why you're paying everyone else first. You may suddenly find a more attractive payment plan made available to you.

Consider also that if you have a credit line with your bank (If you don't, you really should, it's an excellent management tool), it behooves you to borrow from your credit line in order not to miss a cash discount. At today's rate, a bank line will cost about one percent per month (12 percent annual rate). Why give up a five percent cash discount, if the payment is coming in a little late from your client? That's like throwing away four percent of your money!

A good exercise is for you to set aside the time to call each and every vendor and ask what their prompt-pay discount is. With bigger suppliers, the discount is probably published in their price book. Still, it's a good idea to call and ask. When you do call, always ask this question, "Is anyone else getting better terms?" You may find that additional monies are, in fact, available to you. With smaller suppliers, prompt-pay discounts may only be available to those that request them. We learned this lesson the hard way, when our solid-surface fabricator casually mentioned that he offered 2% 10-day terms. It did not say anything about that on his invoices, so we had been paying him in 30 days since the beginning. We had wasted over $5000 in the three years we had done business together. After you've assembled your list of discounts, post it on the wall directly in front of your bookkeeper, with instructions not to miss any discount opportunity without your prior approval. That list represents cash that you want to keep!

It is also your right, within reason, to take a prompt-pay discount well after the billing date, if it is due to manufacturer error. If backorders or "mis-manufactured" materials have held up your project and kept you from getting paid, you should still take your discount whenever the job is finally made right. Make sure that you document the problem, so that your *partners* can understand the situation. Otherwise, they will assume that you're not paying because you have a cash-flow problem. You also will want to charge back any labor costs you incur while repairing or replacing any defective product. Don't forget travel time, material costs, etc. Our industry has many high-quality, high-integrity manufacturers who have no understanding of the retail end of the business. They'll wonder why they're not being paid when they "only forgot the sink base," while you're left with a job you can't install and can't get paid for. If you're honest and forthright with your suppliers, they should respect your reason for holding up payment and then taking the prompt-pay discount when they finally get it right.

10 NEGOTIATE THE RATE AND THE DATE

Most dealers simply ask their suppliers what their payment terms are. They never think to try and negotiate those terms. If a manufacturer tells you that the terms are 2% 10 days, tell him or her that you would like to do business, but you need 3% 10 days or, perhaps, 2% 15 days. This is your cash that you're negotiating, and you owe it to yourself not to leave any on the table.

A savvy negotiator would ask suppliers what their average receivables are in days. If their payments are received on an average of 25 days, then tie in your discount payment term to something just less than that average. After you think you've arranged your best deal (i.e., 5% 15 days), look your supplier straight in the eye and ask if anyone else is getting better arrangements. You may find that the terms suddenly get even better.

It's never too late to broach the subject with a supplier, even if you have been doing business for years. In fact, if you've been a good account, it should be even easier for you to raise the cash discount or, at least, extend the discount terms for five days or more. Every day that you extend the discount term gives you a better shot at not missing a discount opportunity. It's imperative for your long-term financial success that you work aggressively to find every dollar available to you!

Story time...

We once met a dealer who was having a terrible time with his suppliers. He insisted that they were uncooperative and were causing his business great distress. When he started in business, he worked with five different cabinet manufacturers. Each one serviced a different piece of what this dealer thought his market was. By the time we met with him, his supplier list had doubled and he was buying cabinetry from 10 different places, or, at least he was trying to.

When he called us, he was looking for our recommendations of other cabinetry sources! Something didn't sound right to us, so we agreed to visit him at his business. What we found there was the antithesis of today's concept of partnering. This dealer would have his delivery/warehouse person open each and every box delivered to him by each manufacturer. Sometimes a delivery truck would have two or three kitchens on it and it would take four or five hours to unwrap, inspect and repackage every cabinet coming in through the warehouse door. The first day we visited this dealer, they had one trailer truck backed up to the door for over three hours. Meanwhile, a second trailer truck sat idling in the parking area waiting his turn to undergo the lengthy pack and repack process. Even worse, the dealer's delivery person had to go on a delivery himself, so the two manufacturers' drivers were left waiting until he returned.

We learned later, from one of the manufacturers involved, that this dealer was even more difficult than we had originally thought. This particular supplier's driver had arrived at this dealer mid-afternoon on the Tuesday before Thanksgiving, only to be told it was too late in the day to be unloaded. Needing an empty truck to pick up material to go back to his home state, the driver had no option but to spend the night and come back and unload at the dealer's the next day. After undergoing a lengthy inspection of all the cabinetry, the driver missed his return-haul opportunity, and he missed Thanksgiving with his family. The manufacturer refused to honor orders already placed by the dealer and never shipped to them again.

Now, you might think that this dealer had an obsession about receiving damaged or "mis-manufactured" goods. You might also assume that he had numerous experiences in receiving less-than-quality products. But, in fact, he was looking for damages because he hoped to find them! He would get upset at his driver if he couldn't come up with anything wrong. Needless to say, he wouldn't pay anything toward a delivery that had even the slightest problem. We immediately felt sorry for any manufacturer who was partnered with this dealer.

We told him that he didn't need our recommendations of new suppliers, and that what he really needed was a financial whiz to step in immediately to help get his ever-growing payables under control. Unfortunately, his ego far exceeded his common sense, and he let us go. Two months later, the business closed.

11 NEGOTIATE CO-OP FUNDS

Many manufacturers will tell you that their dealers do not use the co-op funds that are available to them. That's probably because dealers choose to advertise their own name rather than that of their suppliers. Still, co-op funds are a source of cash that you should not overlook.

At the beginning of every year, it's a good practice to sit down with the sales representatives who represent suppliers offering co-op dollars. Rather than accept the co-op program as written, you may want to negotiate alternatives that make more sense to your business growth and your cash flow. For instance, you may say, "This year I want to use my co-op funds to help sponsor a particular charitable event." Not many manufacturers will turn that request down; consequently, you've found a way to do something good for your community and earn some favorable publicity as well. We have used such funds towards a show house built for the benefit of our local symphony orchestra, a television studio set to be used by a public television station and for a fundraising kitchen tour. All of these benefited our community while featuring our products and our design capabilities. You may also want to insist that since you're putting in a new cabinet display, you want your co-op funds available to help accessorize it. Again, that's a hard one for a manufacturer to object to.

The key to an adjusted co-op program is to make the arrangements well before you actually go out and spend the money. We have found that sales representatives usually know how to make any legitimate co-op request work.

Aggressively tapping this resource will allow you to keep more cash turning to your benefit.

12 SMILE WHEN PAYING COMMISSIONS!

Many business owners wear a big frown when it's time to pay out commissions. Often times you can hear them grumble to their salespeople, "You make more than I do." If you've ever felt that way, then the problem is in the way you're paying those commissions.

Many commission plans are based on sales. Now, sales are wonderful things. We need sales to continue growing. More sales mean more profits—sometimes. More sales also mean more chance of errors, more deliveries to handle, more warehouse space, more bills to pay and so forth.

We do not necessarily want to suggest that you limit sales, instead, we suggest that you, the business owner, spend your time finding ways to generate new sales. Your commissioned salespeople, however, should spend their time finding ways to generate profit. It is our feeling that paying commissions based solely on profit is the only way that makes fiscal sense. After all, many businesses have run into trouble when they are overwhelmed with sales. No business, however, has ever drowned in too much profit.

One problem with commission plans based on sales is that the business owner is often the only one to take a hit when gross margins fall short. With a plan based on profit, the salesperson has a vested interest, just like the business owner, to make a profit. Mistakes, "mis-orders" and miscues become a shared experience. Why shouldn't your salespeople lie awake at night worrying about profitability just like you do? Let's look at this example:

Sale	$20,000	Sale	$20,000
Gross Profit @38%	$ 7,600	Gross Profit @38%	$ 7,600
Commission @5% of Sale	$ 1,000	Commission @13% of Profit	$ 988
Let's increase the gross margin:			
Sale	$20,000	Sale	$20,000
Gross Profit @40%	$ 8,000	Gross Profit @40%	$ 8,000
Commission @5% of Sale	$ 1,000	Commission @13.5% of Profit	$ 1,080

In this example, the commission would rise or fall a half point for every two points of increased or decreased margin. It is important that salespeople have an incentive to increase profitability. Then both business owner and salesperson are winners. Just as importantly, when mistakes happen, the salesperson not only understands, but also feels the pinch. The commission rate will be lower, but so will the gross-profit base that it is figured on.

As a small-business person, you are a vertebra in the backbone of America. It is not, however, your duty to create jobs. Your sole duty is to generate profit. Profit, then, creates the jobs.

13 GETTING A NEW LEASE ON LIFE

Sometime ago we read an article by a business consultant suggesting that most commercial tenants over-pay their monthly rent, because many small business renters never take the time to figure out what their square footage of usable space really is. We measured our space and, sure enough, he was right—we were over-paying.

The important thing to remember is that your lease is for renting usable space. Some buildings have 12"-thick cinder-block walls. That's your landlord's space, not yours. Measure your entire space from the inside walls. If, for instance, the furnace is the landlord's responsibility, the furnace-room area is the landlord's space, not yours. Even if you come up with only a few square feet of difference, it could mean a sizable amount of money over time.

If your current lease is for 3500 square feet at $8 per square foot, your monthly obligation is as follows:

3500 x $8 = $28,000 ÷ by 12 months = $2,333 per month

However, if you find that your actual usable space is 3200 square feet, it should look like this:

3200 x $8 = $25,600 ÷ by 12 months = $2,133 per month

In the above example, the difference of $200 per month means $2400 per year. This won't work for everyone, but, if it does, you just paid for your next vacation. In our case, we received a credit towards future rent from our landlord. Either way, it's cash in your pocket, and that's our goal!

14 POSTING PROFITS FROM POSTAGE

One of the most easily overlooked expense areas is your postage costs. Admittedly we're not talking about large amounts of money here. Nevertheless, it's your money! Putting some simple strategies in place will help you hold on to more of your cash.

Many of us, along with our staff, simply guess at the amount of postage required to send things out in the mail. We always err on the side of more because we want the material to get to the customer and not be returned for lack of postage. A simple postage scale would tell us exactly what was needed. Further, a postage meter would allow us to put exactly the right amount on our mailings, not the next closest stamp that we happened to have on hand.

Our bigger savings, however, can be achieved by looking at what we mail out. Many of us notify our customers of their upcoming delivery, for example, by typing a letter, putting it an envelope and mailing it. An attractive postcard would do the job for substantially less. Further, an attractive postcard will stand out from the other mail that your customer is likely to receive that day. Take a look at everything your business mails out. Is there a simpler, more attractive, cheaper way to do it? There probably is and it's your cash we're talking about!

15 LAWYERLY ADVICE

Unfortunately, we all need a lawyer from time to time. Hopefully, dealing with legal bills for your payables is the exception, not the rule. In our business, it seems that we can go a year or two without any sort of legal bills, and then—wham!—we're spending all kinds of money. We all want the best legal representation possible, but there are better ways to pay for it.

First, insist, as much as possible, on phone consultations. Whenever you make an appointment with your attorney, he or she is going to book a certain amount of time to meet with you. Inevitably, you're going to spend part of that time talking about the kids, the weather, baseball—something other than your reason for being there. Your lawyer can afford to engage in friendly conversation, because you're paying for it! It's not that lawyers are unscrupulous, it's just that we've never seen a *friendship* deduction on their bills. Phone conferences tend to be much shorter and directed to the matter at hand. When the conference is over, your attorney should invoice you only for the actual time you spent on the phone. It is also quite easy for you to keep a phone log that you can compare with the invoice when you receive it. If follow-up work is needed on your behalf, don't be hesitant to ask if one of your lawyer's lesser-paid staff members could do much of the work for you. The savings could be in the range of $50 or more per hour.

Thinking through your legal options before you actually engage your attorney could result in saving hundreds (thousands in a really bad year!) of dollars. We're sure that you could find much more enjoyable ways to spend that cash!

16 LEAVE THE DRIVING TO SOMEONE ELSE

Over the years, we have rarely found an instance where it made sense for a kitchen and bathroom dealer to support a delivery truck and driver. Yet, often one of the first things people do when starting a business is go out and buy a delivery truck. One of the first employees hired in a start-up dealership is a driver. Why? Who knows? We did the same thing ourselves.

It is definitely worth the exercise to sit down and figure out what a truck and driver cost you each year. The driver's salary is an obvious expense, but, don't forget taxes, worker's compensation, health insurance, vacations, sick days, etc. If you are financing or leasing the truck, the monthly payments are an easy expense to track. But, remember to include the insurance, oil changes, tires, repairs, tolls, truck lettering, washing, etc. When you add all that up and divide it by the number of kitchens sold yearly, you may be in for a rude awakening. It is an even more aggravating exercise when you think back to how many times your driver had to be kept busy, when there were no deliveries. Finally, consider that the truck (that you're paying all this money to keep up) is not being used, at best, more than 200 hours out of an average 720-hour month. When you factor in the occasional holiday, vacation or sick day, the pain inflicted on your checkbook becomes even more damaging.

In most markets there are a number of delivery services available. With a service like this, you pay only for the actual delivery, no more. Delivery services usually charge a flat rate to a particular zone, plus a specified rate per cabinet. This way you can figure your exact delivery expense before you even sign a sales agreement with your client. You should also be able to negotiate a yearly volume-rebate program with an outside delivery carrier. With an in-house truck and driver, those true delivery costs are lost somewhere in your P&L. Most delivery service personnel are quite used to asking for and getting a COD payment, if that's what you want them to do.

Take a hard look at your trucking expenses. It is likely more prudent to be in the driver's seat of controlling your cash by staying out of the driver's seat of driving your own truck.

17 LETTING SUPPLIERS COLLECT THE CASH

As times have gotten more competitive, an ever-increasing number of items have become difficult for kitchen and bathroom dealers to sell profitably. Dealers are expected to offer installation, appliances, granite countertops, and so on. Yet, these items are often sold at ridiculously low-gross margins. The net effect of these and other low-gross products and services is that they drag down the overall gross-profit margin of your business. The true picture of how well you're doing (or not doing) can be distorted. The dilemma is that if you try to raise your prices on these items, it will make you appear expensive and may jeopardize the sale of other higher-grossing products. If you are giving away the appliances at cost in order to get the cabinet sale, you are not doing yourself a favor. One answer, of course, is to stop selling these items. After all, if your operating expenses are running at 27 percent of sales, you're losing money with anything sold at less than a 27 percent gross. However, if you don't offer these items, customers will not think of you as a full-service operation and may want to shop elsewhere.

One answer to the problem is to *arrange* the sale of low-gross products. Tell your clients that you will arrange for them to buy their granite directly from your stone source. Indicate that you will continue to act on their behalf as long as they use this particular stonecutter whose work you trust and admire. Have your granite supplier make out the quotes in your name with an extra 10 percent gross (x 1.15 markup) added. Give the quotes to the customers so they'll know they are buying directly. They will make all payments to your stone source who will then pass on to you a 10 percent rebate. The net result is that your clients will be pleased that they bought directly. Meanwhile, the gross-margin percentage on your P&L does not get destroyed. And, finally, the rebate shows up on your financial reports as *other income*. No *cost of sales* against it, no *overhead*, just *profit*. Your P&L will look healthier and you'll have a much healthier attitude about handling these formerly give-away items. Take a look at this simplified P&L which assumes that you have done $75,000 (which assumes that you have referred $75,000 from your Traditional Sales) in referral business:

	Traditional Sales	With Referred Sales
Sales	$800,000	$725,000
Cost of Sales	502,500	435,000
Gross Profit	297,500 (37%)	290,000 (40%)
Operating expenses	240,000	240,000
Other income	0	7,500
Net income	57,500	57,500

The same rebate program can be easily set up with an installer, local appliance dealer, lighting source, flooring contractor or any other allied professional. With these items removed from the *sales* and *cost of goods* entries on your P&L, you'll get a better picture of how you're doing with your core items. You will also be less hesitant to turn your back on low-gross items that you will have to place in your showroom and talk about anyway. Further, your statements will look better to a banker, supplier or investor.

18 HIRE YOURSELF WITH A CONTRACT

Owners of small, closely held companies are notorious for the free-wheeling way they compensate themselves. When cash is tight, they're the last one to get a paycheck. When cash is flowing, they buy a boat. This lack of structure may not be a good thing for you or your company.

Having no controls on the way you're compensated does not create a sense of stability in the eyes of your bank, your suppliers or the Internal Revenue Service (IRS). This is especially true if you have incorporated your business. A good solution is to prepare an employment contract between yourself and your company. The contract should be made between the President of XYZ Company and the XYZ Company. If your title is something else then use that title. It might help should the IRS ever come calling that the title on your contract be the same one you use on your business cards and in business correspondence. The contract should be updated yearly and should include such things as salary, automobile expenses, insurance (including health, life and auto), travel and entertainment expenses, educational reimbursements, etc. You also want to give yourself wide latitude on the ability to draw a bonus, using wording such as, "Bonus not to exceed 75 percent of the corporation's yearly net profit."

One of the reasons you are doing this is to demonstrate to any creditor or the IRS that you are a competent, thorough, well-reasoned business manager. The bigger benefit to you, however, is that this exercise will force you to think through the way cash is distributed to you. Take, for instance, tax preparation fees. It would seem reasonable, in most cases, that your employment contract dictate that accountant's fees for preparing your personal tax return are to be included as part of your compensation. After all, you and your company are virtually inseparable tax-wise. Under this scenario, your tax preparation costs are an expense item on your P&L. They, like all expense items, lower your company's profit before taxes are figured. If you have been in the habit of writing a personal check to your accountant to cover your personal tax return fees, you have been giving away cash. Your company will be showing more profit, which is taxable, and you will be drawing after-tax income from your checkbook.

As you work through your employment contract, try always to think through the tax obligations that you will face. Your goal should be to have as many items as possible run through your business checkbook rather than your personal one. It is always a good idea to have a conference with an aggressive, competent accountant, so that you don't do something that will come back to bite you.

It is also a good idea to put in writing, be it in contract form or not, whatever compensation agreements you have with your staff members. It could certainly help you should an employee leave under less-than-friendly circumstances.

Remember, although you have an obligation to pay taxes like everyone else, you should never be expected to pay a penny more than the rules require. An employment contract can be a good tool for optimum management of your cash.

 MAKING TAX TIME LESS TAXING

At the end of the year, many business owners find themselves in shock over their impending tax obligations. We are all so busy making sales and managing projects, that we, somehow, overlook this giant barracuda waiting to gobble up our hard-earned income.

In the kitchen and bathroom business, there are ways to manage the tax crunch better. Let's assume that your tax year ends on December 31. December, like any other month, will have projects in various stages of completion. For instance, let's say you're wrapping up a $20,000 job. The supplier invoices have all come in and you've received payment from your client for everything except $100 that they have withheld because some knobs were missing. You have options in how you would like to record this job for year-end.

If you're trying to minimize your tax load in this particular year, you may want to go ahead and pay all of your suppliers in December, but not book the sale until it is fully completed in January. Or, if you're trying to appear more profitable, you might want to book a sale of $19,900 in December and wait to pay your suppliers until January.

Over the long run, you will pay the same amount of taxes either way. As a management exercise, however, the way you handle your year-end closing can have significant benefit to you. If you're thinking of seeking financing, it would behoove you to show as much profit as possible on your tax return. Let's face it, any lender is going to demand to see your return.

Conversely, if at the end of the year, you find yourself in a cash-flow squeeze, you will want to make your tax obligation that year as small as possible. When you consider how many jobs you have going in your year-end closing month, the power you have to produce the results you want are immense. It is definitely worth the time to sit down with your accountant and work on how to present the numbers in your best interest.

Consider this...

I grew up 19 miles south of Boston. In my youth, it was an all-day event to leave town and head for Boston. There was just one main route and it seemed to have 300 traffic lights between our town and the city. On a good day it would take us well over an hour to make the journey.

Consequently, most of our shopping was done on Main Street in our town. I remember well buying all of my clothes and school supplies at the local 5 and 10 cents store. It was owned by a gentleman named Mr. Howard, who was called "Howie" by everyone in town.

I recall visiting Howie's store to purchase some back-to-school sneakers when I was entering 6th or 7th grade. As Howie was measuring my foot, I remember my mother asking him if he was worried about the new superhighway that was about to open, connecting our town to Boston. At each of the two exits off the highway, a discount department store was being constructed. Howie said he wasn't worried at all. He said something about how the level of service he provided far exceeded anything offered by discount department stores. As confirmation of his point, he held up the contraption he had just measured my foot with. I felt better knowing that Howie wasn't going to go out of business.

Six months later a J.M. Fields store opened off one exit and a GEM store opened off the other. The selection in these stores seemed unbelievable and, if, somehow, you couldn't find what you wanted, you could head for Boston and be there in 20 minutes!

A few years later a Bradlees Department Store opened across town. Bradlees was bigger and better, and almost immediately caused the GEM store to give up and close. A few more years went by and a K-Mart moved into the empty GEM store building. Not long after K-Mart opened, J.M. Fields found itself forced out of business. After all, how could they compete with K-Mart? K-Mart had done the unthinkable and had grown bigger than Sears! More years passed and a new store opened in the J.M. Fields' former building. It was called Wal-Mart, and it immediately became the "department store of choice." Bradlees closed its doors soon after Wal-Mart opened and K-Mart struggled along for a year or two before holding its big going-out-of-business sale.

Whatever happened to Howie and his customer service, you ask? His store was boarded up eight months after the highway opened. What has all this got to do with the kitchen and bathroom industry? Well, we believe that most kitchen and bathroom dealers are right now just about where Howie was when he boasted of his high level of customer service. Our industry is only beginning to experience the emergence of the retail behemoths so prevalent in other industries. Kitchen and bathroom dealers have decisions to make. Do you want your business to grow into a kitchen and bathroom superstore? If not, you really have only one other choice. And that choice is to live and breathe quality.

Quality is the only thing that can outsell selection and low prices. Quality design. Quality service. Quality products. All these things add up to being quality business people. The big guys will always be raising the number of the services they offer, and you'll need to raise yours along with them. Work smart, never be complacent and you'll be a competitor. Every day, take a long look at yourself in the mirror and make sure you don't look like Howie!

20 LOOKING YOUR BEST

Many entrepreneurs, on their way to an important bank appointment, think that looking their best involves some extra time in front of the mirror. Unfortunately, most bankers don't care how good looking you are. They're more concerned with how good looking your numbers are.

It is important to identify and understand exactly who wants to see your financial reports and why they need to see them. Your job is to present your numbers to fit your audience. While it is never a good idea to falsify your financial reports, it can be an equally bad idea to use the same format in every financial presentation.

A bank, for instance, wants to see how much cash a business generates. A supplier creditor will probably want to see if you're profitable. Both of these parties can usually be satisfied with the same presentation format. But, if your spouse should hire a divorce attorney, you would want to use a whole different format. You would probably present the IRS with the same financial reports as you would to your spouse's attorney. Delivery is everything!

Here are two different presentations of the same numbers generated by Acme Kitchen:

Acme Kitchens, Inc. Income Statement - 6 months Period 1/1/96 - 6/30/96			
Net Sales	$555,819	100.00%	$555,819
Cost Of Sales			
Product Repairs/Warranty	432	.08	432
Repair Contract Labor	112	.02	112
Repair Product	1,264	.23	1,264
Installation Costs	2,662	.48	2,662
Supplier Purchases	360,230	64.01	360,230
Purchases - Warranty & Repair	54	.01	54
Purchase Discounts	[660]	[.12]	[660]
Freight Out	3,510	.63	3,510
Total Cost Of Sales	367,604	66.13	367,604
Gross Profit	188,215	33.86	188,215
Operating Expenses	180,164		*110,550
Net Income Before Other Items	8,051		77,665
Interest Income	98		98
Interest Expense	5,550		5,550
Rental Income	2,850		2,850
Net Income (Loss)	5,449		75,063

Does not include owner wages and fringe benefits.

Everything was exactly the same until we got to *operating expenses*. There we pulled out the *owner's wages* and *benefits*. Why? To demonstrate the ability of Acme Kitchens to generate cash. If a supplier were to ask, "How much do you make?," a savvy small-business owner would probably say, "I would be glad to provide you with backup information substantiating the numbers presented; however, I regard my personal income as a private matter." Usually, the issue gets dropped there and the dealer gets the needed credit line. To demonstrate as much cash flow as possible, make sure to capture as many owner benefits as possible. Don't forget payroll taxes, insurance, auto expenses, phone, postage, travel and entertainment, etc. Of course, your *statement of operating expenses* would be adjusted in the same manner.

This same exercise of removing the costs of the owner can be the basis for developing a cash value for your business. Almost all small-business valuations are based on the cash flow a new owner would be likely to enjoy should he or she decide to step into your shoes. In fact, a small business with a profitable bottom line is often worth less than a small business with substantial cash flow.

When it comes to arranging your best credit terms or when it's time to sell your business, understanding presentation options could mean a large amount of cash for you. Even if you're looking to do neither, knowing how to make a cash-flow presentation will clearly show you how much income you're generating for yourself. You may find that you can't afford to sell out!

21 FINANCING FOR PROFIT

Ask most kitchen and bathroom dealers about consumer financing and they will tell you that their customers don't need financing. Yet, these same dealers can recall many instances when their clients really wanted custom cabinetry, yet chose a lesser-priced option because it was more in line with their budget. Financing in our industry does not have to mean stickers on our displays saying, "only $49 per month." Although there are many dealers successfully operating this way, there are many more who feel that their business would be demeaned by selling on a so-much per month basis.

Offering financing can be handled in a subtle way. Perhaps a small brochure included with your job proposal or a tag line on your quote saying "financing available" is sufficient. Financing does not have to mean that you're boosting the sale to a point the customer can hardly afford. Instead, financing may offer a nervous homeowner a sense of security that he or she won't run out of money before the project is completed. It can also be to the customer's benefit to finance the project with a tax-deductible home-equity loan, rather than cashing in an interest-earning investment. Some customers may find it less expensive to finance their project than it would be to pay capital gains taxes on cashed-in investments. There are a number of situations where financing could be of great benefit to the consumer. What we're more interested in, however, is making the point that financing can benefit you.

Let's say that over the course of a year you were able to capture one additional $15,000 project because financing was available. Along with that, it may be a fair assumption to calculate that you could upgrade two or three customers to products costing $3000 more per job. These four financed situations would generate an additional $24,000 in sales. With a gross margin of 40 percent, your dealership would realize $9,600 in gross-profit dollars. It is hard to think of a reason that you would not enjoy the additional capital!

22 YOUR SIGNATURE IS WORTH MONEY!

When kitchen and bathroom dealers first start out, they usually do all of their own bookkeeping. They make sure that all of their customer's needs are taken care of before they worry about such things as paying bills and balancing the checkbook. Most often, these tasks are tended to late at night when even the heartiest entrepreneur has trouble thinking straight. Consequently, as the business grows, the first responsibility most dealers assign to others is the job of handling the financial end of the business. After all, most of us went into business for ourselves because we were great salespeople or top-notch designers. We've never met anyone in our industry who started their business because they wanted to be a bookkeeper!

The potential, however, to give away cash is as strong a possibility with your bookkeeper as it is with a designer who can't read a tape measure. There is no doubt that you should be in the habit of paying your bills on a timely basis. Do not, however, pay a bill just because it's due. Pay it because it's right! How do you make sure it's right? By signing the checks yourself!

There is no question that your time is more valuable out on the sales floor or behind a drafting board. Let your office personnel handle all of the other non-income producing tasks. You don't need to open the mail, review purchase orders or log invoices for payment. You should, however, make sure that every check being sent out in the mail has your signature on it. With every check ready to be signed, insist that the gross profit made on the items being paid for be noted along with the check. If you thought that you should be grossing 40 percent on your stainless sinks and the note attached to the check shows 33 percent, then you've got a problem. If the margin isn't what you think it should be, give it back to your office manager with instructions to find out why. Of course, a sharp office manager will already have determined where the problem lies.

In our business, we use "blue slips" (we print these forms on blue paper to make them stand out) for cost discrepancy reports. Our sales and design staff hate to see them coming! Our "blue slips" look like this:

COST DISCREPANCY REPORT

Job Name _____ *Date* _____

Item in Question _____

Item Cost on Purchase Order $ _____

Item Cost on Vendor Invoice $ _____

Difference Between Anticipated Cost
and Amount Invoiced $ _____

Explanation/Action To Be Taken _____

Signature _____

Maybe your staff is forgetting to add freight when they're figuring costs. Perhaps your supplier's representative hasn't updated your price catalog since their last price increase. Your new salesperson may be lowering prices to get orders rather than selling effectively. Your bookkeeper may be getting lazy in not checking to see that the discount on the invoice was correct. Whatever the reason, it's cash that you need to capture and it may be the only way you're going to know it's happening. It's awfully hard to identify where the leaks are if your only method of financial management is to review your monthly P&L. The *margin-busters* will get lost somewhere in your *cost of sales*.

If you cut down the number of times you send out checks, the process will be much easier. By only sending out checks on Friday, for instance, you will only have to review the checks once a week. The time spent Thursday evening or Friday morning will be well worth your while. You may also find that you're needlessly missing prompt-pay discounts. We all work so hard to get money in the front door that we should make it a priority not to let it slip out the back.

23 JOB-COST OR IT'S MONEY LOST

Not too many years ago, job costing was a daunting task. Matching up invoices, purchase orders and sales slips was all done by hand and involved piles and piles of paper. Today there are a number of software programs that make our lives easier. It is amazing, however, how many kitchen and bathroom dealers do not bother to job cost. It's like walking a tightrope without a net!

Job costing is exactly what it sounds like. It is identifying every cost related to a job, so that you can determine *exactly* what the whole job cost and, therefore, *exactly* what your profit is. Relying solely on your monthly P&L to tell you whether or not you have made money will only mask where your problems are.

One of the most overlooked items in any job costing is the time factor. Time, after all, is money. How much time and money did you spend on the project? How many showroom visits did the client make? How many re-draws were done? Every minute working on the client's project represents an outflow of your cash. Doctors sell their time. Lawyers do, too. As a design expert, isn't your time also valuable? Make sure you develop a system to record hours logged to each project. A daily time record such as the one that follows is an ideal tool to use. If you are not willing to regularly invest time in documenting how you spend your day, then, at least, do yourself a favor by filling in the daily time record enough times so that you can develop a reasonable average of actual time spent on your projects. You may be surprised at how much of yourself you are giving to your clients!

DAILY TIME RECORD

	Job Name	Phase Code	Total Time
8:00			
8:15			
8:30			
8:45			
9:00			
9:15			
9:30			
9:45			
10:00			
10:15			
10:30			
10:45			
11:00			
11:15			
11:30			
11:45			
12:00			
12:15			
12:30			
12:45			
1:00			
1:15			
1:30			
1:45			
2:00			
2:15			
2:30			
2:45			
3:00			
3:15			
3:30			
3:45			
4:00			
4:15			
4:30			
4:45			
5:00			
5:15			
5:30			
5:45			
6:00			

Phase Codes: 01 Client Meeting, 02 Estimating, 03 Design, 04 Job-site Measure, 05 Vendor Ordering, 06 Vendor Follow-up, 07 Installation Prep, 08 Installation Monitoring, 09 Service Calls, 10 In-House Meetings

A correctly done job-cost report will show all costs related to a job. Included will be the cost of goods, add-on purchases, replacement costs for mistakes and "mis-orders," time spent, etc. Most importantly, it will tell you what your profit was on each project. An effective manager will then know how to set selling margins on future sales.

Unfortunately, many people in the kitchen and bathroom business are, in fact, not business people. Don't be one of them. Job costing is an important, effective management tool. Make sure to use it!

Story time...

Several years ago we visited a kitchen dealership that was in deep financial trouble. Their bank was threatening to call their loan. Their suppliers would ship only on a COD basis and, even then, would accept only certified checks. The owners insisted that they were only having a temporary cash-flow problem. After all, they said, their monthly financial statements indicated that they were doing just fine. In fact, their financial statements showed that they had been profitable every month since they started in business. They were so confident of their profitability that they had built a big house in an exclusive area, purchased two expensive German cars and had enjoyed some luxurious vacations. The day we arrived at the dealership, a new golf cart was being delivered. What's the big problem?

As with many business problems, the solutions can often be incredibly simple. We all get caught up in our day-to-day affairs which can cause us to overlook even the most basic flaws. Someone from outside our business, however, can often walk in and 10 minutes later correctly diagnose the problem. *That's just what happened on our visit with these folks.*

The former owners of their business had been achieving a 38 percent gross profit and these folks just knew that they were "at least five percent better" salespeople. Adding five percent to the 38 percent, they then directed their bookkeeper to log in all sales showing a 41 percent gross margin. Sure enough, all of their financial statements showed that they had continually achieved 41 percent. We took out some of their purchase orders and began matching them up with payments received on a couple of jobs. It took about an hour to determine that their true gross margin was closer to 27 percent. With operating expenses running at over 30 percent of sales it was amazing that they had managed to stay afloat for their three years in business!

When we explained the problem to them, their first reaction was to tell us that we didn't know what we were talking about. It was impossible for them to believe that they were not just having a temporary cash-flow problem. Eventually they agreed to take the paperwork home and review it for themselves that night. The next day they reluctantly agreed that maybe, just maybe, they had a little problem with the way they were producing their in-house financial reports. Almost casually they asked, "Is this why our tax returns have shown such big losses the last three years?" Aghhh!!

 REIMBURSE YOUR PURSE!

It is often difficult for small-business owners to separate themselves from their business. It is not just the emotional bonding, it's also the financial melding of the affairs of you and your business that you should keep aware of.

When it's time to spend money on business items, many business owners will open their wallets and remark, "It all comes out of the same pocket." That was probably true in the days before taxes were invented, but it is certainly not the case now. Never lose sight of the fact that every dollar you take out of your pocket or personal bank account has already been taxed. If your company is paying you $50,000 each year, consider the fact that tax authorities are going to claim somewhere in the neighborhood of 30 percent of it. That's $15,000 that you're contributing to the common good. Of the 70 cents on the dollar you have left, **do not** spend it on business affairs.

What we're talking about is tolls, gas, entertainment expenses, business travel, postage, long-distance phone calls, personal computers bought for business use, measuring tapes, pens, pencils, and so on. It has always amazed us at how many kitchen and bathroom dealers intertwine their family checkbooks with their business checkbook! It is important that you develop a petty cash reimbursal system. Keep your receipts and, if they're legitimate business expenses, reimburse yourself. Your business, after all, pays for its expenses with pre-tax money. Your personal cash is all post-tax money. That means that the dollars your business has are worth 30 percent more than yours. Those are the dollars that you want to use. It's like getting a raise!

25 WHO ARE YOU, ANYWAY?

One of the great questions raging through our industry is, "Who are we?" Are we retailers or are we design studios? Can a retailer also be a design studio? Can a designer be a retailer?

It's an important distinction that could have a great effect on your bottom line. The large home-center chains have had tremendous impact on the way kitchens and bathrooms are sold. They are open virtually all hours of the day—everyday. They can design a kitchen, finance it and load it in a customer's truck all within an hour. If you think that you're a retailer, then there's your competition. You'll have to figure out how to stay open just as long and how to price your product just as competitively. Every day the home centers get better at what they do and they raise the benchmark that you're going to need to live by. If you are confident that your business is, in fact, a design business, be aware that you're still going to have to react to the way home centers are coming to market. You need to convince potential clients that your design capabilities and service level are worthy of their trust and their investment.

In a design business your expertise and, therefore, your time is a big part of what you are selling. Much like a lawyer, accountant or physician, you are being sought for counsel. If you exude professionalism and do a professional job, you will gain referrals from your clients. It is those referrals, rather than strictly price-selling, that will separate you from them. Being a true professional also allows you the opportunity to structure your business to your benefit.

When you visit your doctor or lawyer, you always make an appointment. On your end, you want to make sure that the professional will be available to see you. On their end, they want to organize their day to best suit their needs. Shouldn't you be doing the same thing, or at least attempting to? Think of the efficiency that comes with scheduled appointments. Rarely would you be short-staffed. Rarely would you be paying overtime. Rarely would you not be ready for your client when they came in to see you. Rarely would you be frantic. Now, we've been in this business a long time and we still have not figured out how to run it like a dentist's office. But we're trying!

One thing we have noticed about other professionals is that they are usually closed two days a week. Making the decision that we were not retailers, we closed our showroom on Sunday and Monday. The effect on our bottom line was immediate. No longer were we rotating our employees' days off. We no longer needed a part-time person to prevent us from being short-handed on any day. Our customers were no longer frustrated when trying to reach the designer they were working with. No longer was one staff member interrupted from what had to get done in order to cover for a colleague that was off that day. In effect, we were finally at full staff every day that we were open. Monday had always been a busy day for us on the phones, so our office people still come in and handle the phones, deliveries, etc. The rest of us use the day to get our own life organized. Doctor, dentist and school appointments are all scheduled on Mondays.

Believing that you're a professional designer means running your business like a professional. Let the retailers go on retailing while you structure your business to best suit your customer and yourself. Making a commitment to professionalism allows you the opportunity to put efficiencies in place that will improve your bottom line. The woman that gives us our haircuts always schedules in a lunch hour for herself each day. If only we were smart enough to figure out how to do that!

26 SERVICE AS MARKETING

If you make the decision that you are, in fact, a design business, you'll want to consider just where your clients come from. If your customer base comes mostly from referrals, then you'll want to make sure that each of your existing clients is absolutely thrilled with the service that you provide. Instead of an expensive marketing program, you may want to consider spending that money on someone whose job it will be to make every customer happy. Having a customer-service manager on your staff could be the best ad campaign you've ever developed!

Many kitchen- and bathroom-design firms do an excellent job for the first 95 percent of the project. Toward the end of the project, the remaining details become increasingly minor and everyone in the firm starts shifting their focus to the next project. Ultimately, referrals are usually won or lost depending on how well that last five percent is handled. Realizing that makes it imperative that each project be left on a positive note.

Some of your customer-service manager's salary can be covered by charging back vendors for time and expenses to repair or adjust "mis-manufactured" items. The rest can be covered by monies originally budgeted for advertising.

At our business, we felt that we were adding a tremendous selling feature by having a customer-service manager on our staff. So much so that we instructed our staff to include a customer-service fee when pricing all of our jobs. We chose to add a fee of three percent to the total job cost before we figure our selling markups. This exercise increases our margins to what we think our service warrants and has yet to cause us to lose a project because of price. Just as importantly, having to figure the customer-service fee on each job reminds our staff that our firm is totally dedicated to complete customer satisfaction.

Having fully satisfied customers is your only guarantee of a steady flow of referrals for your firm. Let's face it, you're going to perform follow-up service for your clients anyway. Why not set it up to your maximum advantage? Put a system in place to make sure that that last five percent of the job goes smoothly. Anytime a client is referred to you by a past customer, the less likely you are to have to adjust your price due to competitive pressure. And certainly, along with offering better service, there is nothing wrong with increasing your cash intake!

27 TO FEE OR NOT TO FEE

No matter what gathering of kitchen and bathroom dealers you might attend, the subject of design fees will always come up. Do you charge? What do you charge? Do you apply the whole amount towards the sale? All are questions you will hear if you put three or more dealers together in the same room.

If your business is set up so that the customer can get a simple design and then drive around to the warehouse to pick up their cabinets on the same day, then you probably don't need to worry about getting a design fee. If, however, you're going to do some serious design work for a customer, then you should really consider what a design fee can do for you and your bottom line.

Calling it a design fee is often a misnomer. Many kitchen and bathroom professionals actually receive design retainers which are applied towards the customer's eventual purchase. Retainers do not cost a client any more money over the scope of the project. Design fees usually are a separate charge added to the customer's total costs. If you view yourself as a design professional, then you owe it to yourself and your cash flow to charge either a fee or a retainer before you impart all of your expertise to your potential client.

Just think of the efficiencies. With a retainer in hand, your chances of landing the job have skyrocketed. With a retainer, you know it's worth your valuable time to draw and price the project. With a retainer in hand, you can unleash the talents (and expense) of your staff on the client's behalf. Without a retainer, you're shooting in the dark.

Most dealers agree that charging a design fee or retainer is the correct way to do business. The hard part for many, however, is standing before the client and asking for it. Yet, if you can't sell a $500 retainer, how are you going to sell the $20,000 job? Even if you did lose two projects a year (which you really shouldn't!) because of customer reluctance to pay the retainer, you still will come out ahead. Just think of all the plans and pricing that you've done only to have the customer buy somewhere else. If you had a retainer in hand, that customer would have had a vested interest in buying from you. You would be able to spend more time maximizing the purchases made by the buyer in hand. You also would be under far less pressure to cut your margins to win the project. The key here, as we said earlier in this book, is to learn to sell yourself not your products. The client needs to be convinced of your expertise and commitment to outstanding service. If you believe in your own professionalism, you will have no trouble in selling a retainer.

Managing a profitable kitchen and bathroom business is not only about charging higher margins and working longer hours. It is also necessary to be efficient with your resources. A design retainer will help you get there.

28 MAKE SURE THE PRICE IS RIGHT

The exercise of pricing a project can be fraught with more danger than anything else you do. If a mistake is made in pricing, it's often difficult to get those dollars back. What can be more frustrating than realizing that you've lost profit dollars even before the project gets underway?

It's a good idea to have everyone in your firm use a standardized price sheet when figuring jobs. That way everyone will be able to figure out what each other did when figuring the costs. Your price sheet should include a space for the customer's name and the date the pricing was done. It should also include a space for the supplier's name, the door style and so forth. On our pricing sheets, we also include spaces for wood up-charges, paint charges, freight-in, delivery costs out, finished-end charges, door and drawer counts, etc.

The idea is to remind the person doing the pricing of all the variables that must be considered in properly pricing a project. It is also a good idea to number each entry on your pricing worksheet so that a corresponding number can be assigned, i.e., to each cabinet on your floor plan. This is a good way to control omissions.

After you've itemized your costs, what price should be put on the various items? You have two options:

1.   Price the item based on what you think the market will bear.

2.   Price the item based on what your sale's ability has earned.

Needless to say, the second option should provide you with more profit dollars. If you are truly offering service to your clients, then they are buying you and your expertise more than they are buying a particular product. Knowing that should help you increase your margins.

Another must is for the owner/sales manager to set the selling prices. Develop a cost/sell sheet for your products. On it, list the names of your suppliers with the selling multipliers that your staff is to use. You can also list the multiplier to find cost as well as the freight factor to get the product to your business. You can make it a rule that no one sells below the pricing as shown, without approval from his or her supervisor. You have now given yourself at least a fighting chance to maintain the gross margin you're trying to reach.

Even better, if you keep your pricing template on your computer, you can make pricing changes monthly, weekly or even daily, if you want to. If your business is swamped with work, isn't that a good time to slide your margins up a point or two? Absolutely! Pricing for maximum profit is what it's all about. The more profit that you can generate, the healthier your business will be. The healthier the business, the better off your employees and customers will be. It is certainly worthwhile to develop a pricing system that promotes accuracy and accountability. Not to do so is like throwing cash out the window.

29 FREIGHT FRIGHT

Freight in to your business and freight out to your customer are two of the most overlooked factors in determining true job costs. Freight charges from cabinet manufacturers are usually easily determined before the sale is made. Other items, however, can have more ambiguous freight costs.

Many distributors will list their prices as FOB wherever they are. That means that they don't want to be responsible for estimating freight costs. You, of course, need to make them responsible. Savvy dealers have said to their suppliers, "We really would like to work with you and sell your products, but we'll need to have some standardized freight costs before we do." It is also wise to carefully check your supplier's invoices for freight costs. We remember one supplier that listed "freight free" on his product price sheets. He offered three deliveries a week for our convenience. What we didn't realize, until we had wasted hundreds of dollars, was that he charged us a $25 "stop" charge every time he delivered material to our place of business. We could have easily bulked our orders if we had been managing our freight costs efficiently.

Bulking freight will almost always realize savings. Try posting your orders right by your fax machine. Take hardware, for instance. If you make it a rule that every Friday your hardware orders get faxed in, then you will have bulked orders all week. When the hardware arrives to you, it will be consolidated in a single container and your delivery-service charges will be minimized. Of course, the fewer vendors you choose to *partner* with, the easier it's going to be for you to get a handle on these costs. The dollars saved may not seem significant individually, but, over the course of a year, they can grow into a sizable chunk of money. Your money! It is also a good practice to specify the week you want delivery of sinks, faucets, accessory items, etc., so that they will coincide with your specified ship date of cabinetry. That way the supplier's invoice will arrive at about the time you're receiving payment from your customer.

The same is true with your freight costs out. In your original pricing, you may have assumed one delivery to a job site. But how many times do you really have to send material out to a site? It's a hard question to answer when you're figuring prices because, at that time, you don't know if everything is going to arrive on time and in good condition. If the whole project goes as planned (which rarely happens!), and you only have to make one extra delivery to replace a damaged door, what will that cost you? Maybe $50? Who is going to absorb that charge? Probably you! Since none of us have a crystal ball to predict how a job is going to go, we suggest that you at least double your anticipated delivery costs when you're putting your proposal together. If you've got a job-costing system in place, you'll be able to develop an average true cost of delivery which you can then factor into upcoming projects. It's an important item to consider, otherwise you'll be taken for a ride with freight costs.

Now, wait a minute...

Over the years we have attended many National Kitchen & Bath Association (NKBA) chapter meetings. It is at these events that we get to know our industry colleagues and, of course, our competitors. One such competitor comes to mind every now and then.

This particular gentleman owned a nice showroom not more than an hour away from ours. He would see us come into the meetings and immediately bend our ear about all sorts of things. I am sure he felt some kinship because our businesses were of a similar type. He told us over and over again that his secret of success was in having his own installers on the payroll. He was adamant that his two full-time carpenters constituted an effective sales tool. He felt that he was controlling projects better and, therefore, was achieving greater success in producing profits. He talked about it so much that we considered putting our own full-time installers on the payroll.

One day, as we were signing a sales agreement with some customers, we asked the clients why they had chosen to do business with our firm. They cited the usual reasons such as design, service, etc. And then they said, "Besides, the other firm we visited said they wouldn't be available to do the installation for at least five months." Five months? That prompted us to start asking all of our clients, from this particular area of the state, if they had visited our competitor there. Sure enough, most had, and many went on to explain that our competitor was too "backed up" to handle their projects. We started to wonder just how many remodeling projects and installations this two-man team could logistically handle in a year? If our competitor was consistently turning away business because his installers were already committed, how could he be making ends meet? We didn't worry about it a whole lot. We were thankful that we had the opportunity to service a bunch of his potential clients.

At one point we were over in his area and decided to drop into his showroom to say hello and to take a look around. He gave us a tour of his facility which ended in his storeroom. There he introduced us to his two carpenters who were busy painting the storeroom walls. We couldn't help but ask why they were painting the storeroom instead of being out on site somewhere. He replied, "I have to find something to keep them busy between jobs."

You probably know where this story is going. About two years after we first met this fellow he called and told us he was going out of business. He could not figure out how to make ends meet. His sales were down and he was behind on his payroll taxes. The reason for his call? He wanted to know if we would like to hire his two installers.

Now, we certainly don't mean to imply that having installers on the payroll is a bad thing. There are some businesses that thrive with installers. In this particular case, however, the way the installers were used actually stifled profits rather than producing them. The amazing thing is that our competitor's problem seemed so obvious to us yet was so undetectable to him.

We can't afford to be too critical. Let's face it, we've all done some pretty dumb things with our businesses. As our industry becomes more sophisticated, however, it will become even more essential for all of us to constantly reexamine why and how we do things. The turnaround time between success and failure gets shorter and quicker every day. If your storeroom needs a coat of paint, you're probably on the right track!

30 INSTALLING INSTALLATION PROFITS

A recent survey done by NKBA indicates that 80 percent of kitchen and bathroom dealers offer some kind of installation services. Some have installers on staff while others prefer to work on a subcontract basis. Either way, the potential for a cash crash is big!

Installation has not historically been a high-margin item for dealers. Yet, the opportunity to be wiped out by *margin-busters* is as high, if not higher, than in any other part of the overall project. One of the more common *margin busters* occurs when the installer needs more crown molding, toe-kick, fillers, etc., in order to finish the job. Another is when the installer bills extra for items not clearly detailed on the plan. Either of these can destroy any hope of profitability on the installation portion of the project. Just as important, the job may get delayed while additional material is ordered and all of your hard-earned trust with the client will quickly wither away. Along with it can go your next job, if a referral is not to be gained.

The only way to protect *your* funds is to insist on an installation meeting between designer and installer before the installation price is given to the client. At that time, every part of the installation should be discussed, including such things as how much crown needs to be purchased and so forth. In our business we have made up a rubber stamp that says, "Installation Check," which must appear on all plans before any cabinetry is ordered. After the installation is underway, if any items need to be procured that are not billable to the client, we adjust the installer's pay or the designer's commission accordingly.

You also want to consider the lead-installer concept developed by NKBA and staunchly professed by remodeling guru, Walt Stoeppelwerth. The concept takes a sharp turn away from the layers of installation and job-site supervision traditionally used. Full responsibility for installation profits are shifted from you to one lead installer. You start by establishing a budget for all costs related to the job. Your lead person should not be paid by the hour, but for the tasks completed, and on the profitability of the entire project. Consider an additional incentive for completion of the project on time and within budget. Under this scenario, there is obvious motivation for your installer. Greater efficiency of time creates greater profit.

A lead installer remains on the job from start to finish. He becomes the continuity clients are looking for on site. Being familiar with all aspects of the job allows him or her to problem-solve as issues come up. Being there each day also provides the communication that your customers are going to need during often stressful installations. Additionally, having that lead installer means that you don't have to stop what you're doing to handle every crisis and answer every question.

The lead-installer concept is an attempt to create total control of job costs. And it sure is nice to have someone besides yourself having a vested interest in producing profits!

Everyone talks with their installers about courtesy to the customers, acting professional, having a neat appearance, being on time, etc. It is just as important to also discuss profitability and the responsibility to produce it. If you don't, the consequences will be like having an open spigot attached to your wallet.

31

SCRUTINIZE AND UNITIZE FOR PROFIT

How many times in your business career have you looked up the price of a cutlery divider? Or, even worse, how many times have you forgotten to include the price of that cutlery tray in your price estimate? The same dilemma exists when estimating labor costs on your installed sales. Did you figure the cost of trips back and forth to the site? How about time spent on the phone? Did you account for each tube of caulk that was used?

Many kitchen and bathroom dealers still use the "stick method" of estimating their projects. They try to estimate every 2" x 4", every piece of plywood, every hour that they anticipate will be spent on site. Unfortunately, it's just not possible to remember everything. Inevitably, pricing by this method will result in a myriad of "nickel and dime" price overages by the time the project is complete. Is there a better way? We think so.

It makes sense to develop some flat rates for parts of projects that happen over and over again. You could establish flat-rate pricing for sink disconnects, sink installs, electrical outlets, per cabinet installation, etc. If you look at these items individually, you'll be able to figure out an average of what and how long it takes to complete the repetitive parts of projects.

Take, for example, installing a laminate countertop. If you study your next installation, you'll be able to determine how many screws you need, how much caulk and glue it took, how many person hours were spent, how long it took to get back and forth to the site. When you have an accurate cost on that job, you can then break it down by running foot or by square foot. The next time you estimate this part of a kitchen or bathroom project, you won't have to start the estimating process all over again. More importantly, you will be less likely to suffer from future *margin-busters*, because they will have been covered by your unit price.

Development of standard costs leads to a flat labor rate for standard elements of installation. Unit costs also lead to quicker and more accurate estimates. This whole method of *unit cost estimating* is best explained in NKBA's Installation Manual. NKBA is also developing a "unit cost estimator" to help you bring the flat-rate pricing method into your business.

We also recommend a pricing checklist to make certain that you have, indeed, included all the items that your customer expects. It makes sense to review the checklist with your customer at the time you sign your sales agreement. Then there will be no question as to whether a cutlery tray was to be included. For you, it will provide one more opportunity to capture the cost of that item before you sign your name.

Eliminating pricing omissions translates into more profit. More profit should mean more cash in your pocket. And more cash in your pocket is our goal, right?

32 CASH-FLOW UNIVERSITY

A common lament among kitchen and bathroom dealers is that the profit they thought they were going to realize on a job often disappears by the time the project is done. There is no doubt that ours is a complicated business. From the time we sign the sales agreement until we collect the final payment there are countless starving *margin-busters* looking to devour our profits. Some of them are probably sitting at desks right in your office!

Many small-business owners complain that their staff doesn't understand that their actions can seriously affect the bottom line of the company. Worse yet, owners often feel that their employees just don't care. After all, employees are paid for their time, including the time it took them to make a mistake. They're paid again for their time while they try to fix their mistake. In between, they spend your money rectifying the problem they created. We've all had employees who didn't worry much about their employer's fiscal health. Most, however, will care if you explain to them how it all works.

Consider that you may be working on kitchen projects that are larger than what your staff member earns in a year. To your employee that is a huge amount of money. It's not hard to see why your staff may assume that there is plenty of cash available to handle whatever minor errors they may make. It's up to you to show them just how critical every dollar is to the well being of the business and to their jobs.

A good exercise is to bring your staff together and follow a project through from beginning to end. If you're job costing, it should not be difficult to identify the items that make up your cost of goods. Identifying operating costs in relation to a specific job, however, is something that your staff has probably never considered. If your firm did six projects in a given month, divide each of your operating expenses by six to develop an average cost per project. Employees (and owners!) will often develop a whole new appreciation of things like heat, electricity, phone bills, advertising costs and so forth, once they understand the impact they have on each job's profitability. If you subtract each project's share of overhead from the gross profit, your staff will start to understand the significance that each of these items has. That huge sale you started with will be reduced to a surprisingly (for them— hopefully not you) small amount of net profit.

If you take the time and do it right, your presentation of the numbers should inspire comments like, "We could have done that job with fewer deliveries," "There's no way that job should have warranted all those phone costs," "Do we turn the thermostat down at night?" With your whole staff present, you'll find staff members examining each other's role in respect to profitability. There is nothing wrong with a salesperson having to explain to peers why costs ended up so high on the project.

Of course, to make this exercise really work, you're going to have to offer some incentive. You should break it down to a simple format. For example, if you want to make 40 percent gross profit (or tie it to net profit) on a job, reward staff members involved with that project $100, if the goal is achieved. Offer nothing if the gross is 39.9 percent or less. If a project falls short of your target, show them the numbers so they can see where they went wrong. When you set a definitive goal, most employees will want to achieve it. As a business manager, if you can get your gross margin up two points on a project, the $100 will be money well spent. For your staff members, over the course of a year those $100 bonuses could represent an additional month's pay.

It has been our experience that most employees want to do a good job. Unfortunately, they are often not allowed to feel part of the process. Take the time to share with them exactly how money enters and exits the business. The result of your efforts should be a happier, more efficient staff with the additional benefit of increased capital in your checkbook.

Meeting for at least a few minutes each week to go over the financial reports will help your staff see the connection between what they do and how it affects the company's well being. Just as importantly, you'll dispel any rumors that the owner is hoarding cash in some secret vault!

33 PLANNING FOR "CASH OUT"

FROM DAY ONE

One of the best things you can possibly do for yourself is to plan right from the start for your exit from your company. It is not something that should be left for "when that day comes." Unless you've figured out how to use a crystal ball, you have no idea what tomorrow may bring. It may deliver great sadness or, hopefully, great joy. Either way, you want to be ready. Besides, planning for your eventual "cash out" is an excellent management tool.

Regardless of whether you plan to sell your business or not, the value of the business can only be ascertained by determining what someone should pay for it. Of course, what someone *should* pay and what someone *will* pay are often two different things. Nevertheless, understanding the market value of your business could have great impact on the well being of you and your family. Your goal every day, starting with your first day of business, should be to have an understanding of the value of your enterprise.

Who is to say how one's health will be a year from now? Will your partnership break up? Who knows what other opportunities may present themselves? Will McDonald's want to build their new restaurant on your site? Maybe someone will come knocking on the door asking if they could buy the business—it happened to us.

The value of your business is important for estate planning. How can you determine what will happen to you in your mature years without knowing the value of your business? Bankers will want to know. The IRS will want to know. A divorce attorney will want to know. Basically, anyone you owe money to will want to know. Most of all, you should want to know.

Understanding value will give you something to work for. We all have days when we want to lock the showroom door and never come back. If you had some idea of the value you were creating, it could help temper your frustrations. It could help you feel that there was, in fact, meaning to your life!

There are many books available to help you put a price on your business. It will be far more beneficial to you, however, to have someone outside of your business do the valuation. That way you'll be able to eliminate your emotional attachment to the business from the valuation process. There are many qualified business brokers who would perform this service for you. Don't hesitate on this one. You owe it to yourself.

Understanding what brings value to your business will guide you in the day-to-day managing of your business. If you approach every business decision with the idea that you may have to explain it to someone in the future, you will most likely make better and more reasoned decisions. For instance, it would be nice to have a new truck, but will it add value to the business? Will a new owner of your business appreciate the money invested in this truck? Will financing or leasing this truck have a positive or negative impact on the value of the business? Will paying for this truck affect cash flow so much that the business loses value? All are important questions that you might not ask yourself if you are not conditioned to think about business value.

We all need rewards in life. Your greatest reward as a business owner could come the day you exit the business on your own terms. That won't happen unless you understand what your hard work has built. Of all the ways you can maximize cash for yourself and your business, understanding and building business value will have the greatest return. Not to plan for that eventual "cash out" starting right now could squander a significant portion of your future financial security. Don't let it happen to you!

Conclusion...

There you have it—33 ways to maximize cash flow for you and your business. Some of these may seem elementary to you. Others may represent a whole new way for you to look at your business. We sincerely hope that you have found at least one strategy that will help you hold on to more of your hard-earned cash.

Ours can be a difficult business. It is fraught with potential margin-busters. They can sneak up and take a bite of profits without you even knowing that its happening. Hopefully, we've helped you to know where to look for some of these insidious creatures.

Our industry is poised for great change in the near future. Who knows how we all will be doing business in five years? One thing for certain is that only the fiscally astute kitchen and bathroom dealers will prosper. We plan on being one of them. We hope you do, too.

NKBA BUSINESS RESOURCES

MANAGING YOUR KITCHEN AND BATHROOM FIRM'S FINANCES FOR PROFIT, Don Quigley
ISBN 1-887127-10-0

LEVERAGING DESIGN: Finance and the Kitchen and Bathroom Specialist, Debi Bach
ISBN 1-887127-31-3

HOW TO INCREASE YOUR KITCHEN AND BATH BUSINESS BY 25%...Starting Next Week, Bob Popyk
ISBN 1-887127-33-X

PROVEN PROMOTIONS FOR KITCHEN & BATH BUSINESSES, Jim Krengel and Lori Jo Krengel
ISBN 1-887127-05-4

BRINGING TOTAL QUALITY MANAGEMENT TO YOUR KITCHEN AND BATHROOM BUSINESS, David Newton
ISBN 1-887127-03-8

NUTS & BOLTS RESOURCES
(For people who work too many hours to find time to read!)

- **Magazines**

  ### INC. MAGAZINE
  Subscription Service Department
  POB 51534
  Boulder, CO 80321-1534
  > One of the best small-business magazines around! Filled with real-life stories of business start-ups, successes and failures. An invaluable resource filled with inspiring and innovative ideas on how to grow a business.

  ### INDEPENDENT BUSINESS MAGAZINE
  IB America's Small Business Magazine
  Group IV Communications, Inc.
  125 Auburn Court #100
  Thousand Oaks, CA 91362
  805 496-6156
  > An excellent cut-and-copy magazine chock full of resources in the *IB* Follow-Up File at the end of each feature article. *IB* Follow-Up File resources include books, software, products, etc., complete with addresses, prices and phone numbers for purchase. A genuine time-saver for available resources on any small business topic that interests you!
  >
  > Article topics range from: *Business Bloopers* (The 10 Most Common Mistakes and How to Avoid Them) to *Peak Performance* (10 Tips for Keeping Yourself and Your Employees Motivated).

- **Books**

  301 GREAT MANAGEMENT IDEAS FROM AMERICA'S MOST INNOVATIVE SMALL COMPANIES, Introduction by Tom Peters, Edited by Sara P. Noble. Copyright 1991, Goldhirsh Group, Inc. ISBN 0-9626146-4-5 (Paper)
  > A compilation of hands-on management ideas based on the practical wisdom of hundreds of real small businesses. This book is presented with humor in an easy-to-flip-through format.

  CASH IN ON CASH FLOW: 50 Tough as Nails Ideas for Revitalizing Your Business, A. David Silver. Copyright 1994, AMACOM, a division of American Management Association, 135 W 50 St, New York, NY 10020. ISBN 0-8144-0210-0
  > Serious information that examines business expense categories and identifies strategies to determine the difference between core and peripheral products of a business. The bottom line is to refocus on the core business and generate cash flow.

- **Organizations**

  NFIB
  National Federation of Independent Businesses
  Membership: 1-800-NFIB NOW
  > NFIB represents the interests of small and independent business owners before federal and state legislative and executive branches of government. An excellent business resource to keep abreast of political and economic issues that impact you and your business.